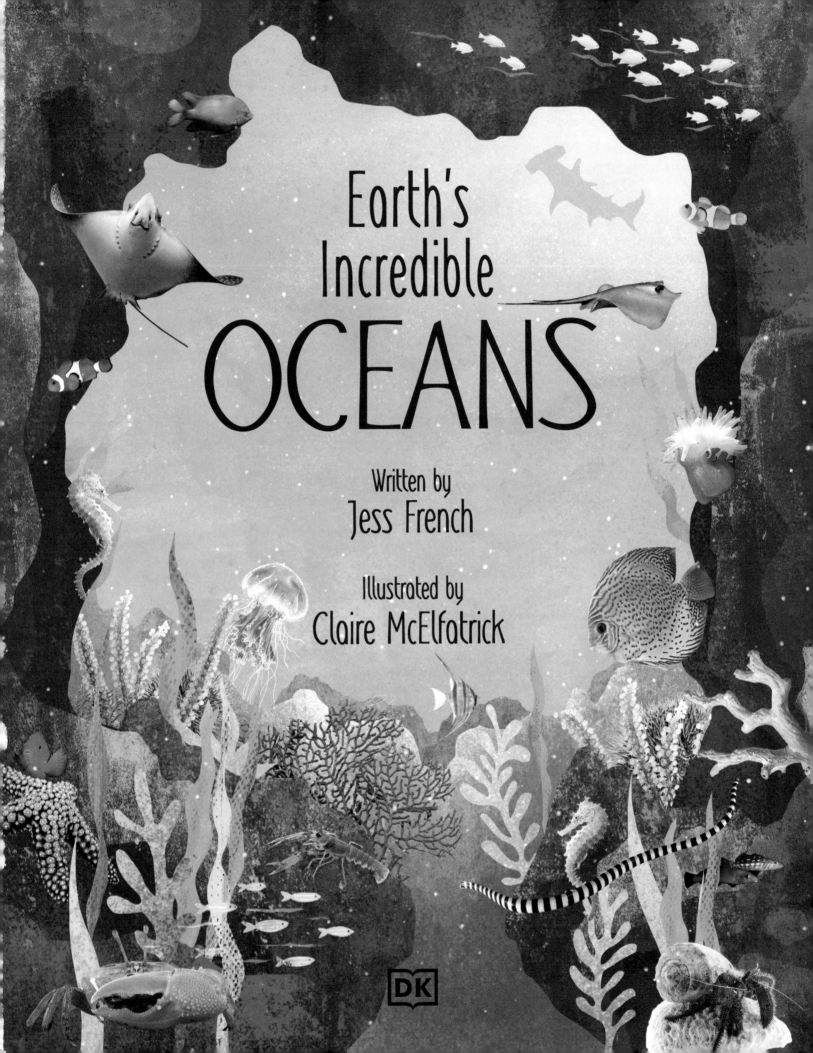

Earth's Incredible OCEANS

Written by
Jess French

Illustrated by
Claire McElfatrick

DK

DK | Penguin Random House

Author Jess French
Illustrator Claire McElfatrick
Senior Art Editor Claire Patane
Senior Editor Lizzie Munsey
Project Editor Clare Lloyd
Design Assistance Rachael Hare, Charlotte Milner
Subject Consultant Helen Scales
Educational Consultant Jenny Lane-Smith
Production Editor Abigail Maxwell
Producer John Casey
Jacket Designer Claire Patane
Jacket Co-ordinator Isobel Walsh
Picture Researcher Sakshi Saluja
Managing Editor Penny Smith
Managing Art Editor Mabel Chan
Creative Director Helen Senior
Publishing Director Sarah Larter

First published in Great Britain in 2021 by
Dorling Kindersley Limited
One Embassy Gardens, 8 Viaduct Garden,
London, SW11 7BW

Copyright © 2021 Dorling Kindersley Limited
A Penguin Random House Company
10 9 8 7 6 5 4 3 2 1
001–321025–Apr/2021

All rights reserved.
No part of this publication may be reproduced, stored in or
introduced into a retrieval system, or transmitted, in any form,
or by any means (electronic, mechanical, photocopying,
recording, or otherwise), without the prior written
permission of the copyright owner.

A CIP catalogue record for this book
is available from the British Library.
ISBN: 978-0-2414-5914-0

Printed and bound in China

For the curious
www.dk.com

INTRODUCTION

Growing up at the coast, I have always felt a connection to the ocean. I have seen such incredible things there, on the boundary between land and water. Yet the life and geography I experienced at the seaside was only a tiny glimpse into the amazing ocean world. From deep-sea yetis to vibrant coral reefs, the ocean is bursting with bizarre and beautiful sights. Come with me as we dive headfirst into Earth's incredible oceans and uncover the secrets of their strange and mysterious depths.

Jess French

Jess French

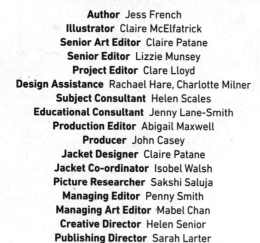

CONTENTS

WHAT IS AN OCEAN?

The Earth is covered in so much water that from space our planet looks blue.

This water is called the ocean. It covers most of the Earth's surface and is crucial to life as we know it. The ocean is so vast and deep that large areas of it are still unexplored. Unlike the water in rivers and lakes, ocean water is salty.

Ocean water is constantly moving. It can be blown by the wind, pulled by the gravity of the Moon, or pushed around by an earthquake.

Read on to learn more about Earth's incredible oceans.

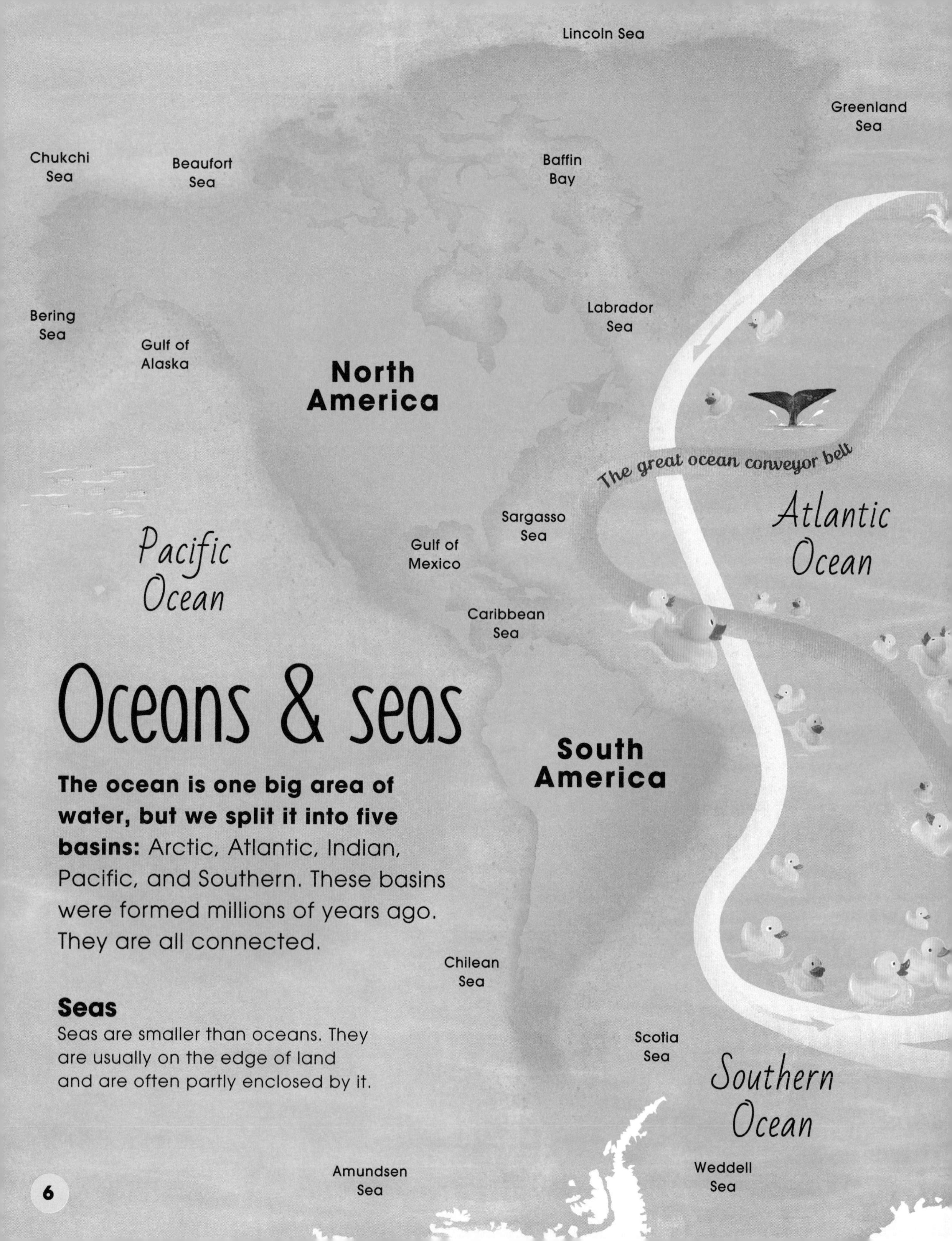

Lincoln Sea

Greenland Sea

Chukchi Sea

Beaufort Sea

Baffin Bay

North America

Bering Sea

Gulf of Alaska

Labrador Sea

The great ocean conveyor belt

Atlantic Ocean

Pacific Ocean

Sargasso Sea

Gulf of Mexico

Caribbean Sea

Oceans & seas

The ocean is one big area of water, but we split it into five basins: Arctic, Atlantic, Indian, Pacific, and Southern. These basins were formed millions of years ago. They are all connected.

South America

Chilean Sea

Seas

Seas are smaller than oceans. They are usually on the edge of land and are often partly enclosed by it.

Scotia Sea

Southern Ocean

Amundsen Sea

Weddell Sea

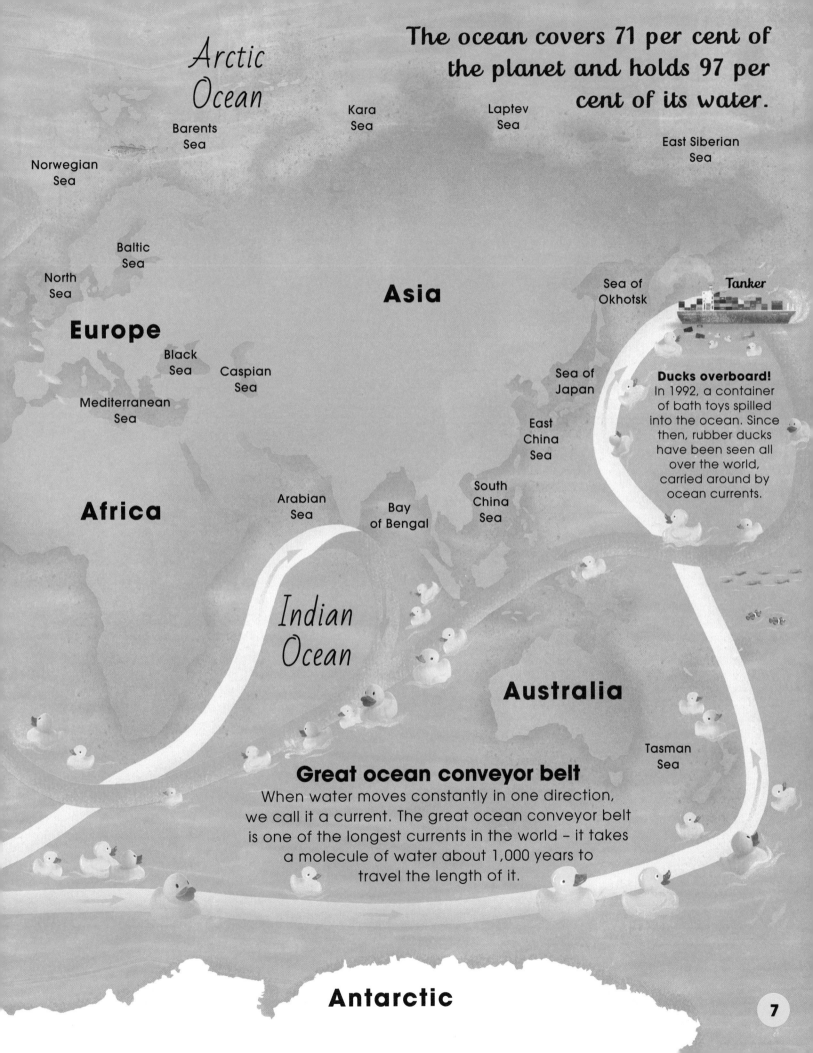

Arctic Ocean

The ocean covers 71 per cent of the planet and holds 97 per cent of its water.

Barents Sea

Kara Sea

Laptev Sea

East Siberian Sea

Norwegian Sea

Baltic Sea

North Sea

Asia

Sea of Okhotsk

Tanker

Europe

Black Sea

Caspian Sea

Sea of Japan

Ducks overboard!
In 1992, a container of bath toys spilled into the ocean. Since then, rubber ducks have been seen all over the world, carried around by ocean currents.

Mediterranean Sea

East China Sea

Africa

Arabian Sea

Bay of Bengal

South China Sea

Indian Ocean

Australia

Tasman Sea

Great ocean conveyor belt
When water moves constantly in one direction, we call it a current. The great ocean conveyor belt is one of the longest currents in the world – it takes a molecule of water about 1,000 years to travel the length of it.

Antarctic

The ocean in motion

Ocean water is always moving, but how it moves depends on many things, including **wind** speed and direction, **water** temperature and depth, and even the actions of the **Moon.**

COLD WATER DOWN

Water that is cold and salty is heavy, so it sinks.

OCEAN CURRENTS

In the upper layers of water, **currents** are created by the wind. In deeper water, currents are driven by the **temperature** of the water and how **salty** it is.

WARM WATER UP

Warmer, less salty water moves up to the surface.

Water is coldest at the North and South poles.

BIG WAVES

When wind blows over the surface of water, **waves** form. When the wind and beach conditions are right, **giant waves** are created.

Wave riders
Surfers travel from all over the world to ride big waves.

TSUNAMIS

When something like an **earthquake** moves a large volume of water, a **tsunami** is created. Tsunamis are huge waves that can travel very quickly and do a lot of damage on land.

Danger waves
Fast-moving waves spread out from the earthquake site.

Earthquake

Tall wave
Tsunamis become higher as they reach shallower waters near the shore.

TIDES

Just like the Earth, the Moon and the Sun have **gravity**, which pulls things towards them. When their gravity pulls on our oceans it changes the sea level, creating **tides**.

High tide

Low tide

Up and down
Tides change in a regular pattern, so we can predict when they will happen.

Sea level

Continental shelf
A shallow area of
ocean floor next
to the land.

Seamounts
Volcanoes on the
ocean floor, with
their tops below
sea level.

Mid-ocean ridge
A mountain range
that runs along the
ocean floor.

Continental slope
A slope that joins the
continental shelf and
the abyssal plain.

Abyssal plain
A flat plain found under
most of the ocean.

The ocean floor

**The bottom of the ocean is one of the most
mysterious places on the planet.** In fact, scientists
know more about the surface of the Moon than the
ocean floor. It has not yet been fully explored, but
scientists have identified some of its key features.

Volcanic island
A volcano that rises up from the ocean floor, with its peak above sea level.

Offshore drilling rig
A platform anchored to the ocean floor, used to extract oil and gas.

OCEAN FLOOR RESOURCES

Many of the things found on or below the ocean floor are useful to people. However, extracting them can be very difficult, especially if they lie in the deeper parts of the ocean.

Guyot
A seamount with a flat top.

Deep-sea canyon
A trench in the continental slope.

Oceanic trench
A deep canyon on the ocean floor. The deepest trenches reach more than 7 miles (11 km) below sea level.

The deepest oceanic trench is the Mariana Trench, in the Pacific Ocean. The world's highest mountain, Mount Everest, could fit inside it with room to spare!

Coral

Seaweed

Jellyfish

Dolphin

School of fish

Shark

Sabertooth fish

Barreleye

Whale

SUNLIGHT ZONE

It is warm and sunny here, with lots of plants. The temperature and light changes depending on the time and season.

200 M
(650 FT)

TWILIGHT ZONE

Only a tiny amount of light reaches this far. The animals that live here have huge eyes to find their way in the gloom.

1,000 M
(3,280 FT)

MIDNIGHT ZONE

No sunlight reaches this zone. It's totally dark, except for glowing light from many of the sea creatures. Without the heat of the Sun, the temperature here is cold and unchanging.

2,000 M
(6,560 FT)

Gulper eel

Wolf eel

Cookiecutter shark

Gossamer worm

Deadly light
Angler fish attract smaller creatures with a glowing lure.

Anglerfish

Sperm whale

Diving deep
Sperm whales can dive as far as the midnight zone in search of food.

Squid

Lancetfish

Animals here must be adapted to life in the cold and dark.

Dead whale

3,000 M
(9,840 FT)

Coffinfish

ABYSSAL ZONE

Very little can survive in the crushing, cold water of the abyss. The few animals that do live here feed mainly on dead plants and animals that fall from above.

4,000 M
(13,120 FT)

The water here is pitch black and nearly freezing.

Magnapinna squid

5,000 M

Brittle stars

HADAL (TRENCH) ZONE

Deep at the bottom of the ocean, dark and mysterious trenches are home to a few extraordinary creatures. We know very little about life down here, as the trenches are difficult to explore.

Cusk eel

Amphipod

Mariana snailfish

Layers of life

The ocean is enormous. It covers more than 70 per cent of the Earth's surface. It also stretches down thousands of metres to the seafloor. Below the surface, the ocean is split into five different zones.

More people have been to the Moon than have explored the deepest trenches of the ocean.

6,000 M (19,680 FT)

7,000 M (22,970 FT)

8,000 M (26,250 FT)

9,000 M (29,530 FT)

10,000 M (32,810 FT)

11,000 M (36,090 FT)

Ambulocetus
(lived 50-48 million
years ago)

Plesiosaur
(200-66 million
years ago)

Ancient oceans

**Billions of years ago, life appeared in
the ocean.** Conditions were very different
to now. At first, only tiny creatures existed.
Since then, many incredible animals have
come and gone, leaving clues about
their lives buried in the sand.

Megalodon
(23-3.6 million
years ago)

Anomalocaris
(540-485 million
years ago)

Cameroceras
(470-425 million
years ago)

Frilled shark
(95 million years
ago-present day)

Pterygotus
(428-372 million
years ago)

Trilobite
(540-250 million
years ago)

INVERTEBRATES

Trilobite

Ocean invertebrates were the first animals. Armoured trilobites filled the seas for millions of years.

FISH

Megalodon jaws

Fish have existed for over 500 million years. The megalodon had jaws and teeth, but the first fish had no jaws or bones.

REPTILES

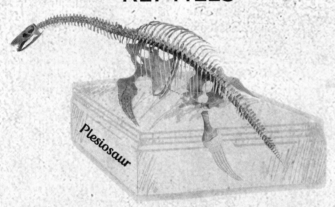

Plesiosaur

Two hundred million years ago, the seas were ruled by huge reptiles with long necks and paddle-shaped feet.

WHALES

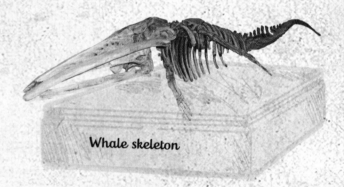

Whale skeleton

Whales evolved from four-legged land animals. Over time, they lost their legs and their arms turned to flippers.

Nautiloid fossil

Nautilus today

Some animals today are almost exactly the same as those that lived millions of years ago.

USNS Bowditch

At the surface
Scuba divers use air-filled cyclinders to breathe underwater.

Exploring the sea
Research ships carry scientists, explorers, and technical equipment.

Exploring the ocean

Unlike fish, humans cannot breathe underwater. We need special equipment to help us investigate the ocean. However, the ocean is vast and some areas are difficult to reach, so even with all this technology, there are still large areas we have not yet explored.

Remote control
Some robots are controlled remotely. Others are designed to find their own way around.

ROV Kiel 6000

Fixed cameras
Sometimes scientists leave cameras in fixed positions, so they can see what happens over time.

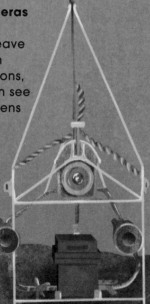
Eye-in-the-Sea

ROBOT EXPLORERS

Robots are used to explore places that are too dangerous for people. They can collect samples and take photos to help scientists on the surface.

EV Nautilus

Under the sea
Ships carrying sonar equipment can send down sound waves to learn more about the ocean.

SONAR

Sonar equipment uses beams of sound to spot objects under the water, communicate with other vessels, navigate through the water, measure depth, and make detailed maps of the seafloor.

Dolphins also use sonar to find things underwater.

Alvin

SUBMERSIBLES

Submersibles are highly specialised watercrafts. They take people underwater to carry out scientific studies.

Bouncing back
The amount of time it takes for sound waves to return to the ship helps build a picture of the ocean below.

Swimming hunter
Polar bears mainly live on land, but will take to the water in search of food.

OCEAN ANIMALS

Earth's oceans are home to millions of different types of animals and plants. They must be specially adapted for a life below the waves.

Ocean animals come in all shapes and sizes, from the tiniest plankton to the most enormous blue whale. Each creature has a job to do in ocean life, and relies on other animals to survive.

Let's dive in and meet some of the fascinating animals that live in our oceans.

Invertebrates

Invertebrates are animals without internal skeletons. They eat bacteria and algae, and are eaten by many larger ocean creatures.

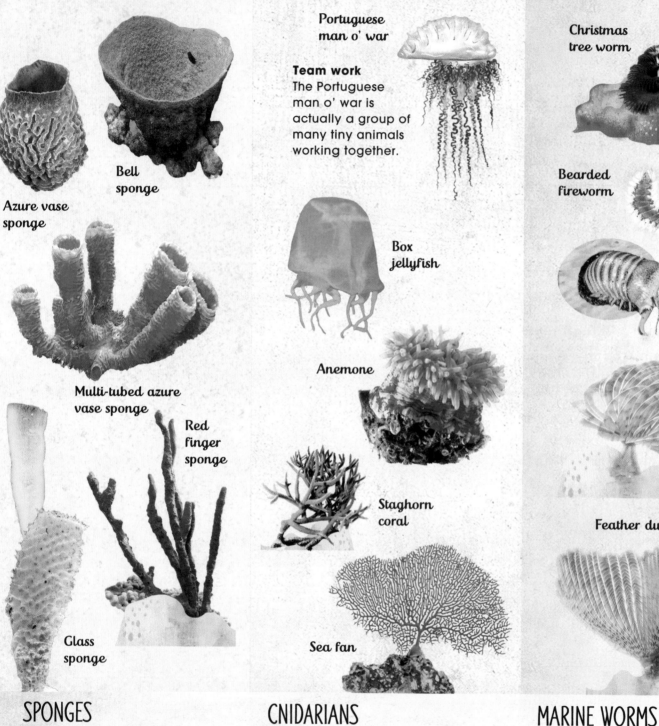

Azure vase sponge

Bell sponge

Multi-tubed azure vase sponge

Red finger sponge

Glass sponge

Portuguese man o' war

Team work
The Portuguese man o' war is actually a group of many tiny animals working together.

Box jellyfish

Anemone

Staghorn coral

Sea fan

Christmas tree worm

Bearded fireworm

Bobbit worm

Feather duster worms

SPONGES

Sponges are very simple animals, but they play a very important role. They eat tiny pieces of waste in the water, which cleans it.

CNIDARIANS

This huge group includes jellyfish, corals, and sea anemones. There are more than 10,000 species of cnidarians.

MARINE WORMS

There are several families of marine worms. Many live under rocks, in mud, or in sand. Often, only a tiny part of the worm is left unburied.

GASTROPODS

This group also contains the slugs and snails that are found on land.

Limpets

Nudibranch

Cone snail

BIVALVES

These animals have a shell in two parts, with their soft body in the middle.

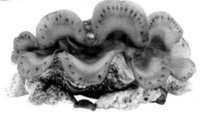

Queen scallop

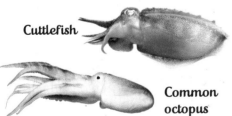

Giant clam

CEPHALOPODS

This group includes octopuses, squid, cuttlefish, and nautiluses.

Caribbean reef octopus

Cuttlefish

Common octopus

MOLLUSCS

There are over 70,000 types of mollusc in the ocean. The animals in this group can look very different. Many of them have hard, protective shells.

Blue spiny lobster

Spider crab

Goose barnacles

Acorn barnacle

Peacock mantis shrimp

Strawberry shrimp

CRUSTACEANS

The hard outer skeletons of crustaceans give them protection from predators, but they must shed their skin in order to grow.

Feather star

Sea star (starfish)

Sand dollar

Sea cucumber

Sea urchin

ECHINODERMS

Echinoderms are only found in the oceans. Their body parts are often arranged in a circle, like the spokes of a bicycle wheel.

Jellyfish

Jellyfish have lived in our oceans for millions of years. They are invertebrates, which means they have no bones in their bodies. These wobbling wonders are known for their stinging tentacles – some have stings that are powerful enough to kill humans.

SIMPLE BODIES

Jellyfish have basic bodies with only a few parts. They are not good swimmers and mostly use water currents to get around.

Bell
This smooth dome is the main part of the jellyfish's body.

Moon jellyfish

Tentacles
Stinging tentacles stun prey before it becomes dinner.

Oral arm
This arm moves food to the jellyfish's mouth.

Life cycle

Most jellyfish start off as tiny larvae, which float down and attach themselves to rocks. They change form several times before becoming an adult "medusa". In some species, dead cells from the medusa can grow into polyps again.

Larva

Polyp

Polyp with buds

Moon jellyfish can get younger as well as older!

GLOWING JELLIES

Some jellyfish make their own light – they appear to glow in the dark. This is called bioluminescence.

A big group of jellyfish is called a bloom. Blooms can contain thousands of jellies.

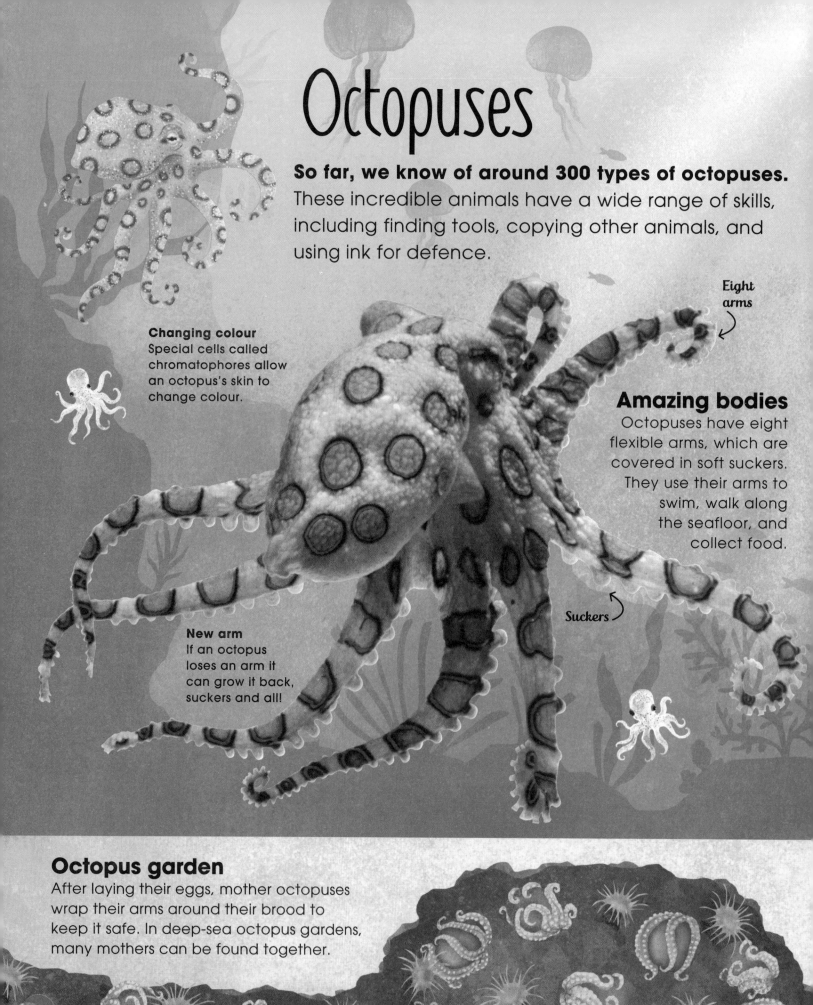

Octopuses

So far, we know of around 300 types of octopuses. These incredible animals have a wide range of skills, including finding tools, copying other animals, and using ink for defence.

Eight arms

Changing colour
Special cells called chromatophores allow an octopus's skin to change colour.

Amazing bodies
Octopuses have eight flexible arms, which are covered in soft suckers. They use their arms to swim, walk along the seafloor, and collect food.

Suckers

New arm
If an octopus loses an arm it can grow it back, suckers and all!

Octopus garden
After laying their eggs, mother octopuses wrap their arms around their brood to keep it safe. In deep-sea octopus gardens, many mothers can be found together.

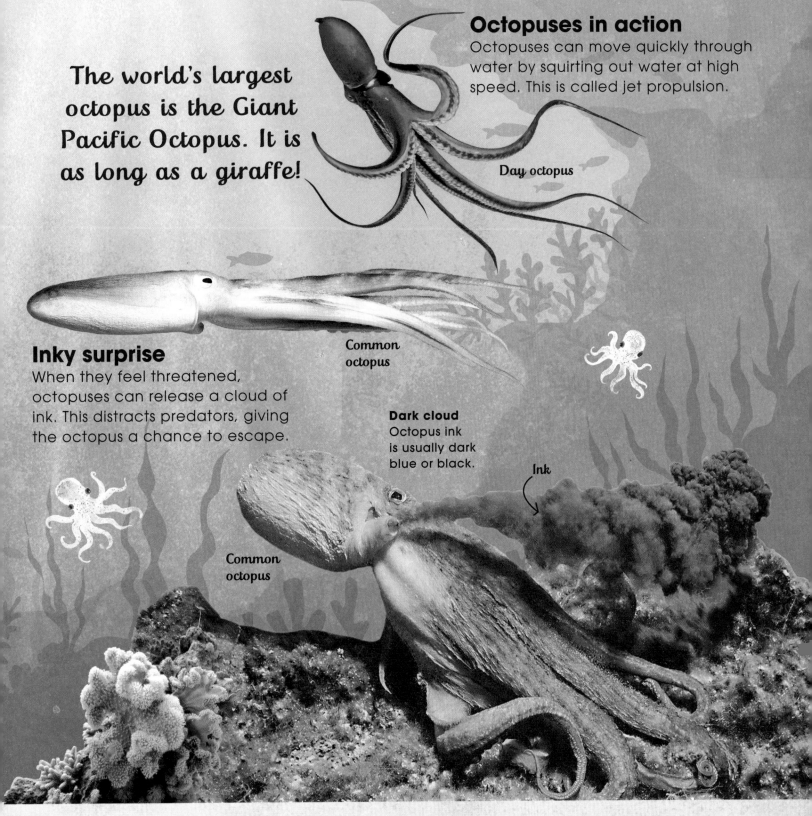

The world's largest octopus is the Giant Pacific Octopus. It is as long as a giraffe!

Octopuses in action
Octopuses can move quickly through water by squirting out water at high speed. This is called jet propulsion.

Day octopus

Common octopus

Inky surprise
When they feel threatened, octopuses can release a cloud of ink. This distracts predators, giving the octopus a chance to escape.

Dark cloud
Octopus ink is usually dark blue or black.

Ink

Common octopus

Coconut collector
Coconut octopuses seek out coconut shells, which they collect to use as helmets, shields, and shelters.

Coconut octopus

Sharks & rays

Sharks, rays, and chimaeras have skeletons that are made of a flexible material called cartilage, not bone. They are also known as the **"cartilaginous fish".** These fascinating and beautiful animals have roamed the seas for over **400 million years!**

Cownose ray

Manta ray

The giant oceanic manta ray has a wingspan of up to 9 metres (29 feet)!

Tail

Pectoral fin

RAYS

Skates and rays have flat bodies, making it easy for them to hide on the ocean floor. There are many different types of rays, including stingrays, electric rays, manta rays, and sawfish.

Mega movers
Manta rays are enormous. They flap their pectoral fins to move themselves through the water.

Cephalic lobe
These 'horns' guide tiny animals called plankton into the manta ray's mouth.

SHARKS

There are hundreds of shark species. Most sharks have super senses, powerful jaws, and sharp teeth. They are incredibly skillful predators.

Hammerhead shark

Gills, for breathing

Tough skin

Several rows of teeth

Bendy skeleton
Shark skeletons are made of strong, flexible cartilage.

Fin

Tail

Speedy swimmers
Streamlined scales, torpedo-shaped heads, and lightweight skeletons allow sharks to shoot through the water at speeds of up to 72 kph (45 mph).

Zebra shark

Australian ghost shark

RATFISH AND GHOST SHARKS
This group are known as the chimaeras. They mostly live in the deep sea, so they have large eyes to help them to see in the dark.

Shark eggs
Many sharks and rays lay their eggs in leathery cases, which they attach to seaweed.

The baby inside the eggcase can take over a year to develop.

When the baby emerges it looks like a tiny version of the adult.

Empty eggcases sometimes wash up on the seashore.

Fish

Most fish are cold-blooded, scaly-skinned animals, which are perfectly adapted for swimming. They are found all around the world, in both salt water and fresh water. Most fish belong to a group called the **"bony fish"**.

Bony fish

This group of fish has a number of special features.

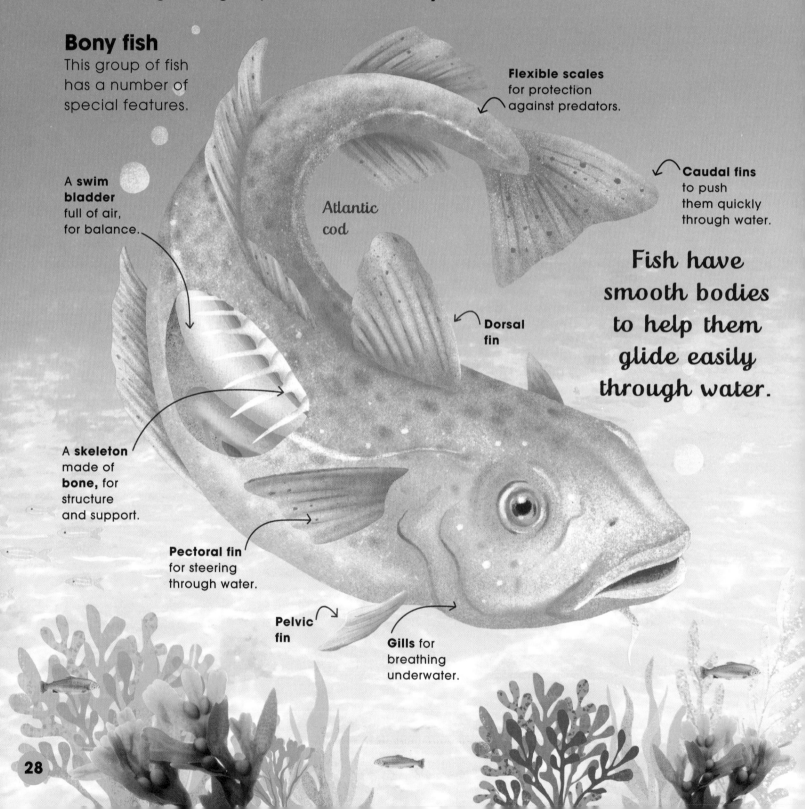

Flexible scales for protection against predators.

Caudal fins to push them quickly through water.

A **swim bladder** full of air, for balance.

Atlantic cod

Fish have smooth bodies to help them glide easily through water.

Dorsal fin

A **skeleton** made of **bone,** for structure and support.

Pectoral fin for steering through water.

Pelvic fin

Gills for breathing underwater.

Normal size

Puffed up

PUFFERFISH

When they feel threatened, pufferfish gulp in water, making their bodies three times bigger.

LIONFISH

The spectacular lionfish is as beautiful as it is deadly. It eats other fish and has poisonous spines.

SEAHORSE

These extraordinary and unusual fish have upright bodies, long snouts, and no scales.

Around 95 per cent of all fish are bony fish. There are over 20,000 different types.

ELECTRIC EEL

Electric eels can make electrical charges, which they use to stun their prey and defend themselves.

FLYING FISH

Flying fish leap out of the water to escape predators, using their huge pectoral "wings" to soar through the air.

GIANT OARFISH

Giant oarfish live deep down in the twilight zone. They are the longest of all the fish.

CATFISH

Catfish often feed at the bottom of the water. They use their long whiskers to find small creatures on the seafloor.

ATLANTIC SALMON

The Atlantic salmon spends half of its life in a river, and half in the ocean. It is also known as the "King of Fish".

Ocean reptiles

Only a few reptiles can survive the salty conditions of the ocean. Of those that enter the waves, only a handful spend their whole lives at sea.

Crocodiles

Several types of crocodile enter the ocean from time to time. The largest of them, the saltwater crocodile, is the best adapted to ocean life. It is usually found near land, but can also ride ocean currents to cross large areas of seawater.

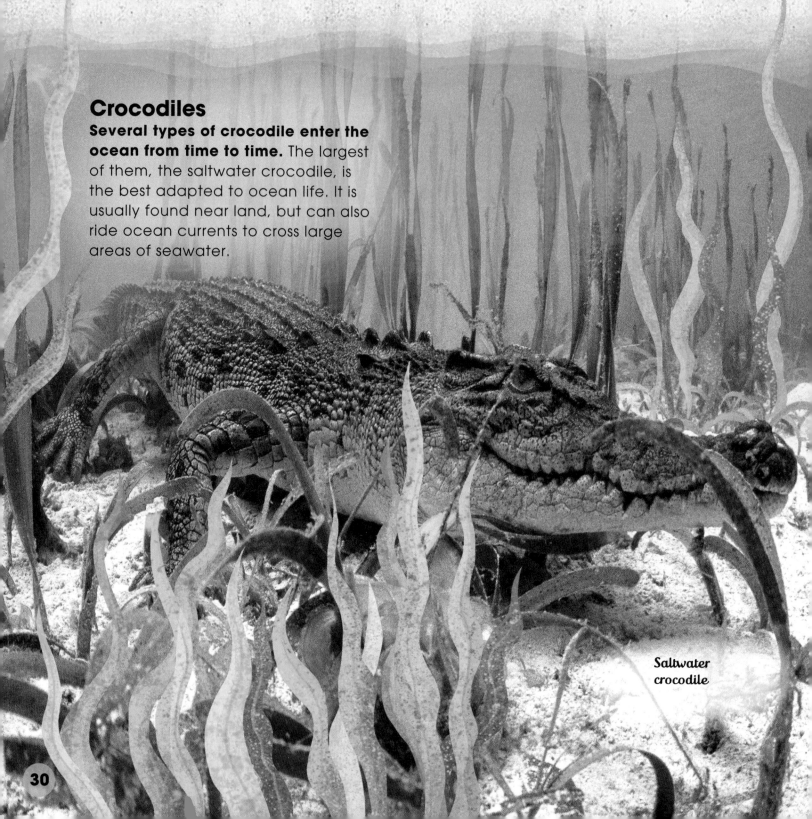

Saltwater crocodile

Yellow-lipped
sea krait

Sea snakes

Some snakes spend their whole lives at sea. They have nostrils on the tops of their heads, and lungs that go most of the way along their bodies to their paddle-shaped tails. Other sea snakes spend part of their lives on land.

Most sea snakes give birth to live young.

Marine iguanas

Marine iguanas are the only lizards that spend time in the ocean. They live in the Galápagos Islands, where they feed on algae growing on the rocky seabed.

Marine iguana

Sea turtles

There are seven types of sea turtle. They are found in oceans all over the world. Turtles spend most of their time in the water and only come out to lie in the sunshine or lay their eggs.

Leatherback
turtle

Out at sea
Unlike a tortoise, a sea turtle cannot pull its head and legs inside its shell.

Seabirds

Life above the ocean can be tough. Birds that spend their lives at sea must drink salty water, find food beneath the waves, and stay in the air until they can find a safe place to land. Some seabirds roam ocean skies for **months on end,** returning to land for only a short time each year to breed.

Light-mantled sooty albatross

ADAPTATIONS

Seabirds are specially **adapted** to survive at sea.

Albatrosses

These are the **largest** of all the seabirds. They can live for over 50 years and travel over 8.5 million km (five million miles) in their lifetimes.

Salt glands to get rid of salt after the bird has drunk seawater.

Long, thin wings to soar for a long time without flapping.

Waterproof feathers

Wandering albatross

Webbed feet for swimming.

Black and white colouring for camouflage in and above the water.

Rockhopper penguins

Penguins

Using their wings as flippers, these expert **swimmers** propel themselves through the water at speeds of up to 35 kph (22 mph).

Auks

Adapted to colder climates, auks have barrel-shaped bodies, small wings, and are excellent **swimmers** and divers.

Tropicbirds

These spectacular tropical seabirds have a loud, **screeching** call and perform impressive flying displays.

White-tailed tropicbird

Atlantic puffin

Skuas

These birds are fierce aerial pirates that **chase** other seabirds to steal their food. They also dive-bomb anything that threatens their nests.

Arctic skua

Petrels, fulmars, and shearwaters

This group are superb **fliers** that spend almost their entire lives on the ocean. They can vomit their stinky stomach oil at their enemies.

White-faced storm petrel

Seabirds often nest in large colonies on beaches and coastal cliffs.

Male frigatebirds have bright-red throat pouches, which they inflate to attract females.

Frigatebirds

Unlike most seabirds, frigatebirds don't have waterproof feathers. They stay in the air for **weeks,** snatching fish and squid as they leap out of the water.

Magnificent frigatebird

33

Ocean mammals

All mammals are warm-blooded, breathe air, and have hair, so the ocean doesn't seem like the most obvious place for them to live. But ocean mammals have some incredible **adaptations to help them live their lives in the water.**

Walrus

Sea lion

Elephant seals

SEALS AND WALRUSES

This group is called the **pinnipeds.** They have front and rear flippers, so they can get around on land and ice, as well as swimming. They often live in cold places and have a thick layer of blubber (fat) to keep them warm.

DOLPHINS, PORPOISES, AND WHALES

These are the **cetaceans.** They have streamlined bodies that allow them to move easily through the water. They also have a special blowhole on the top of their heads, so they can breathe air from the surface.

Spinner dolphin

DUGONGS AND MANATEES

These gentle vegetarian giants are found in warm, shallow waters. They have large, thick bodies and bristly lips for pulling up seagrasses. They are closely related to elephants and it is thought they were once mistaken for mermaids!

Dugong

Manatee

Blue whales are the largest animals ever known to have lived.

Blue whale

Until laws were made to ban whale hunting, many whales were in danger of extinction. Since the ban, some whale populations have started to recover.

Humpback whale

Beluga whale

Narwhals

Famous for their unique spiral tusks, the mysterious narwhals are known as the unicorns of the sea. They are only found in the chilly Arctic, often poking their tusked heads up through holes in the ice.

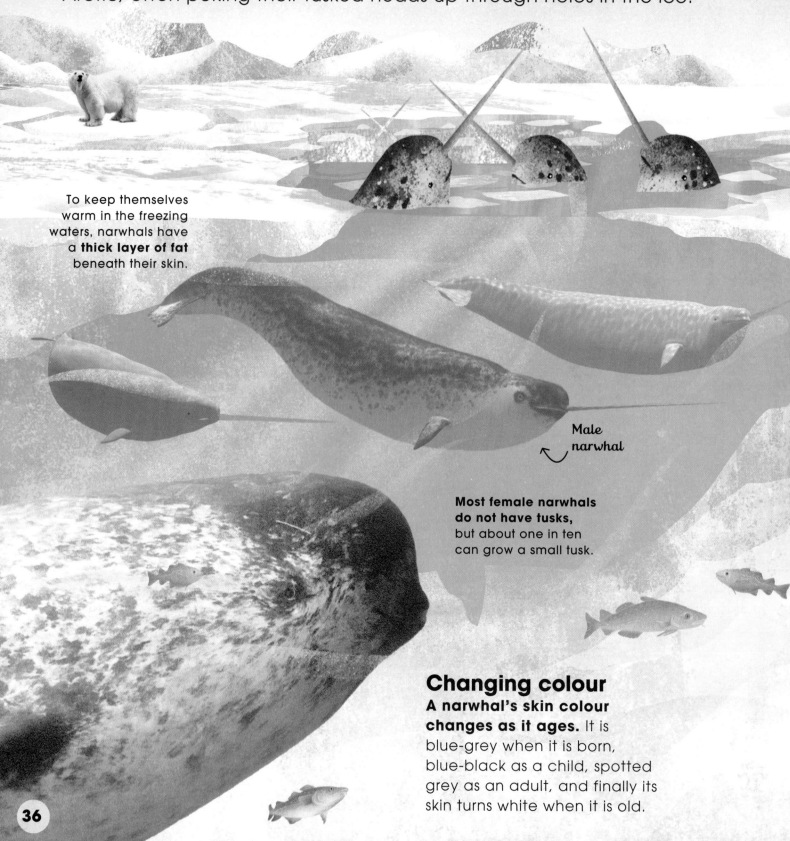

To keep themselves warm in the freezing waters, narwhals have a **thick layer of fat** beneath their skin.

Male narwhal

Most female narwhals do not have tusks, but about one in ten can grow a small tusk.

Changing colour
A narwhal's skin colour changes as it ages. It is blue-grey when it is born, blue-black as a child, spotted grey as an adult, and finally its skin turns white when it is old.

Narwhals travel together in groups, called pods, of up to 100 members.

Arctic tern

Ringed seal

Narwhals poke through gaps in the ice. The narwhal breathes in air through a **blowhole** in the top of its head.

Males may **joust** with their **tusks.**

Fishy food
With only two teeth in its mouth, the narwhal must **suck up food.** It mostly likes to eat squid, fish, shrimp, and crabs.

Arctic cod

The narwhal's tusk is actually an **overgrown tooth.** It only has one other tooth in its mouth.

Narwhals dive underwater to catch their food, holding their breath for up to **25 minutes.**

LIVING IN THE OCEAN

In the ocean there is competition for everything, including food, shelter, protection, and mates. It pays to have some special tricks up your sleeve...

Ocean animals have evolved crafty ways to survive and thrive. Pregnant dads, shrimp cleaner stations, and whale childcare – the ocean is home to fascinating behaviours that are not seen anywhere else on Earth.

Dive deep into this chapter to learn more about life in the ocean.

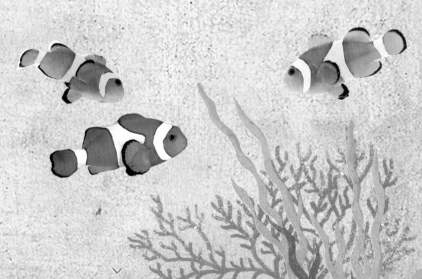

Searching the seashore
Gentoo penguins search for food close to the shore, making hundreds of dives every day.

Food web

In the ocean, everything is connected. Energy is passed from creature to creature through a dense **food web.** In most cases, the food chain begins with **plants and plankton,** which get their energy from the Sun.

Gaper clam

Krill
These tiny animals exist in every ocean. They are eaten by thousands of other animals.

Tufted puffin

Plankton

Krill

Lion's mane jellyfish

Turtle cracking
The powerful jaws and teeth of tiger sharks help them crack through hard turtle shells.

Green turtle

Pacific herring

Regal tang

Blue-ringed octopus

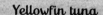
Tiger shark

Yellowfin tuna

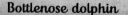
Bottlenose dolphin

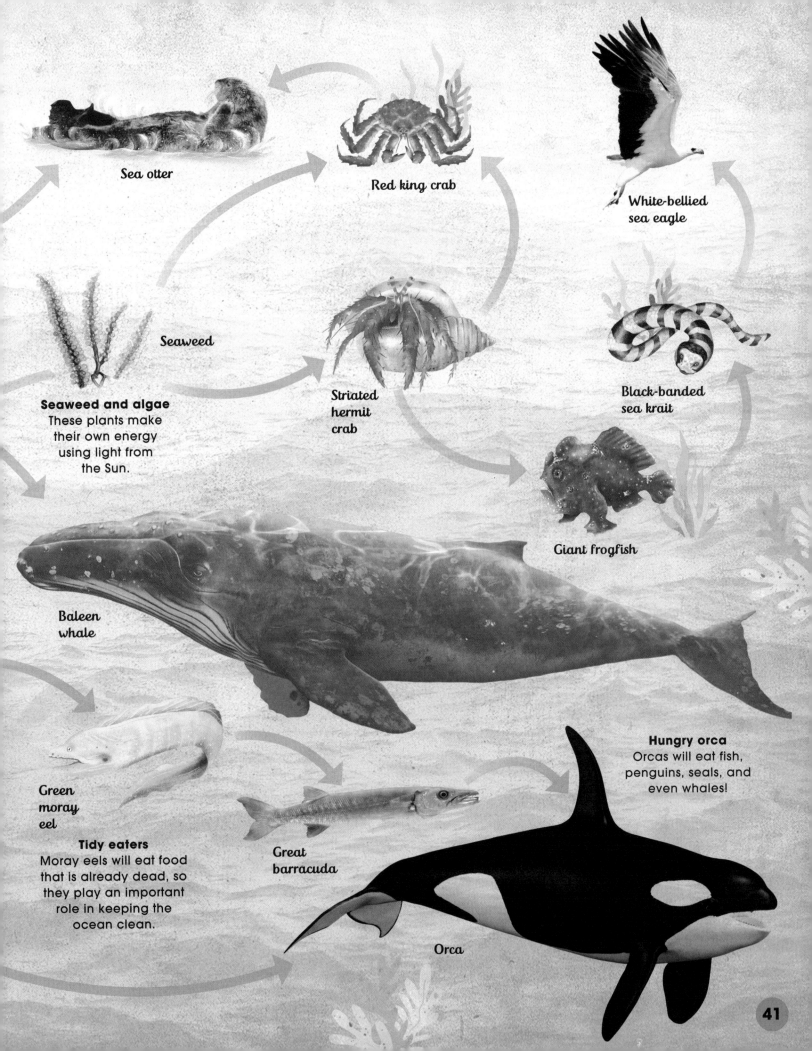

Sea otter

Red king crab

White-bellied sea eagle

Seaweed

Seaweed and algae
These plants make their own energy using light from the Sun.

Striated hermit crab

Black-banded sea krait

Giant frogfish

Baleen whale

Green moray eel

Tidy eaters
Moray eels will eat food that is already dead, so they play an important role in keeping the ocean clean.

Great barracuda

Hungry orca
Orcas will eat fish, penguins, seals, and even whales!

Orca

Hiding in plain sight

With danger behind every wave, ocean animals must find ways to hide from predators who would like to eat them. Some predators also use **camouflage,** allowing them to lie in wait for their prey without being spotted.

MIMIC OCTOPUS

These octopuses are masters of disguise. They twist their bodies to make themselves look like other poisonous and foul-tasting sea creatures.

Boneless bodies
Octopuses have no internal skeletons. This makes it easier for them to change their shapes.

Blurred outline
Tassels break up the outline of the wobbegong, making it harder to spot.

TASSELLED WOBBEGONG

This carpet shark has a beard of wormlike, fleshy tassels lining its upper lip. Small fish may mistake them for food and move closer, but get eaten instead.

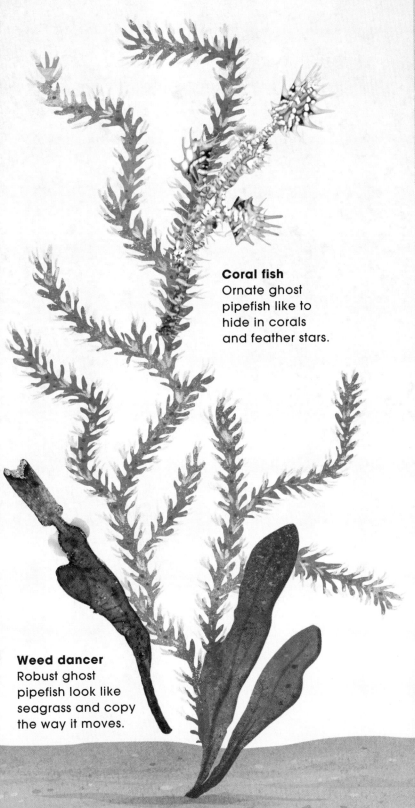

Coral fish
Ornate ghost pipefish like to hide in corals and feather stars.

Weed dancer
Robust ghost pipefish look like seagrass and copy the way it moves.

GHOST PIPEFISH

Ghost pipefish are closely related to seahorses. Their bodies look very similar to plants and corals, allowing them to blend in perfectly with their surroundings.

STONEFISH

The stonefish lies on the reef, waiting for smaller fish to swim past so it can eat them. It has poisonous spines to keep it safe, which are strong enough to kill a human.

Deadly spines
When a stonefish sits still on the floor of a reef it is almost impossible to spot.

Safety in numbers

Many fish stay safe in the ocean by working as a team. Some types of fish, such as tuna, herring, and anchovies, always travel in groups. Others only form groups when they need to – for example when hunting or mating.

WHY DO FISH SWIM TOGETHER?

Swimming in a group is a great way to confuse predators and reduce your chances of being eaten. Working as a team can also make it easier to find food.

SHOAL OR SCHOOL?

Fish travel together in many ways. Groups of fish are given different names depending on how they behave.

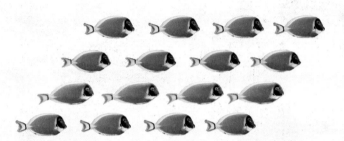

SCHOOL

A school of fish is very organised. All of the fish in a school are the same type, a similar size, and they move together in the same direction.

SHOAL

A shoal of fish is less structured than a school. It may contain many different types of fish and even other animals.

Lateral line

Lining up

Fish use a special organ called the lateral line to help them move with the others around them. The organ feels tiny changes in water pressure that are created as the other fish swim around.

Blacktip
reef shark

Watch out!

This school of fish have all moved together to get away from the shark.

Working together

Ocean creatures help each other out in many ways.
Sometimes, animals of totally different types have relationships that are useful to them both – this is called "symbiosis".

DOLPHIN MUD RINGS

Dolphins are famous for their clever hunting tactics and often work together to catch fish. One way they do this is by creating mud rings to trap their prey.

Unlucky escape
Confused fish leap out of the water and into the mouths of the waiting dolphins.

Fish

Dolphin

A ring of cloudy water is created around a shoal of fish.

Muddied waters
The dolphins beat their tails to bring up mud from the seafloor.

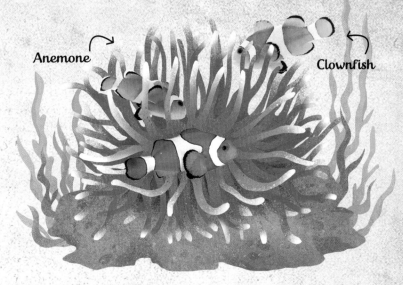

Stinging shelter

Anemones allow clownfish to live safely within their stinging tentacles. In return, the clownfish bring in oxygen-rich water, clean the anemones, and feed them with clownfish poo.

Anemone

Clownfish

Goby

Pistol shrimp

Shared tunnels

The goby is a little fish that lives in a tunnel dug and cleaned by a pistol shrimp. The shrimp can't see very well so when danger approaches, the goby raises the alarm, and both animals dart into the tunnel to hide.

Bluestreak cleaner wrasse

Yellow boxfish

CLEANING STATIONS

In coral reefs, fish and other animals go to certain places to be cleaned. The cleaners get a tasty snack, and the reef is kept healthy and free of disease.

Complete clean
Large fish have their whole body cleaned, including inside the gills and mouth. They don't eat the cleaner fish.

Leopard moray eel

Banded boxer shrimp

Orca family

Female orcas can live for over a hundred years! Grandmother orcas care for their grandchildren, sharing food with their daughters' babies.

Orca with calf

A mother blue whale produces up to 220 litres (50 gallons) of milk for her baby every day – that's enough to fill a bath!

Whale babysitters

When sperm whales dive into deeper waters, they leave their calves at the surface with other mothers. These babysitters will let the calves drink their milk if they are hungry.

Sperm whale with calf

Loving care

In a dangerous environment like the ocean, family is very important. Baby sea creatures can make a tasty snack for ocean predators. Most have a better chance of surviving if there are adults around to protect them.

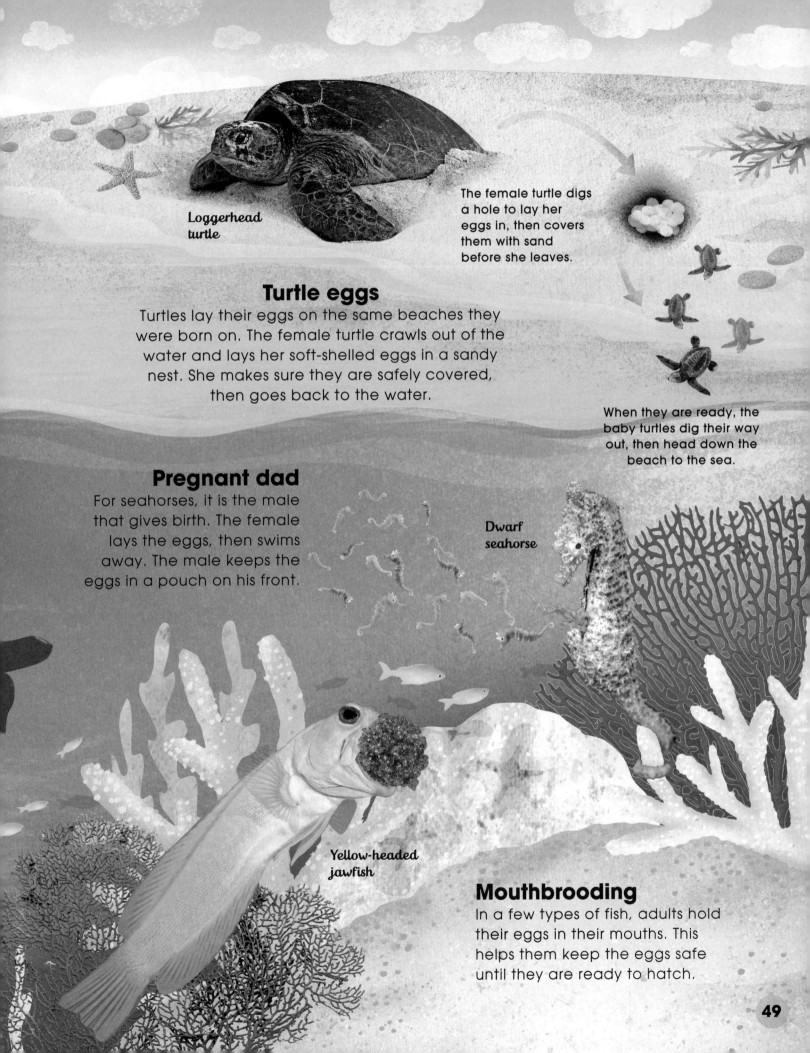

Loggerhead
turtle

The female turtle digs
a hole to lay her
eggs in, then covers
them with sand
before she leaves.

Turtle eggs

Turtles lay their eggs on the same beaches they
were born on. The female turtle crawls out of the
water and lays her soft-shelled eggs in a sandy
nest. She makes sure they are safely covered,
then goes back to the water.

When they are ready, the
baby turtles dig their way
out, then head down the
beach to the sea.

Pregnant dad

For seahorses, it is the male
that gives birth. The female
lays the eggs, then swims
away. The male keeps the
eggs in a pouch on his front.

Dwarf
seahorse

Yellow-headed
jawfish

Mouthbrooding

In a few types of fish, adults hold
their eggs in their mouths. This
helps them keep the eggs safe
until they are ready to hatch.

Animals on the move

Ocean animals don't always stay in one place. They move around in search of food, mates, or places to nest. Some make yearly journeys, travelling long distances across the world. This is called "migration".

WHALES

During the summer, humpback whales feed in the cooler waters by the poles. In the autumn, they move to warmer waters to breed. Calves swim near larger whales to make their journey easier.

TURTLES

Female sea turtles can swim for thousands of miles to reach their nesting beaches. They often return to the beaches they hatched on to lay their own eggs.

Loggerhead
sea turtle

LOBSTERS

In the winter, spiny lobsters march in long lines across the ocean floor. They travel to deeper waters, using the Earth's magnetic field to help them find their way.

Spiny
lobster

At night, some deep-sea creatures swim to shallower waters to feed. They include shrimp, jellyfish, and squid.

Humpback whale

A whale calf is able to swim with its mother from birth.

Spiny lobsters travel across the ocean floor for up to seven days.

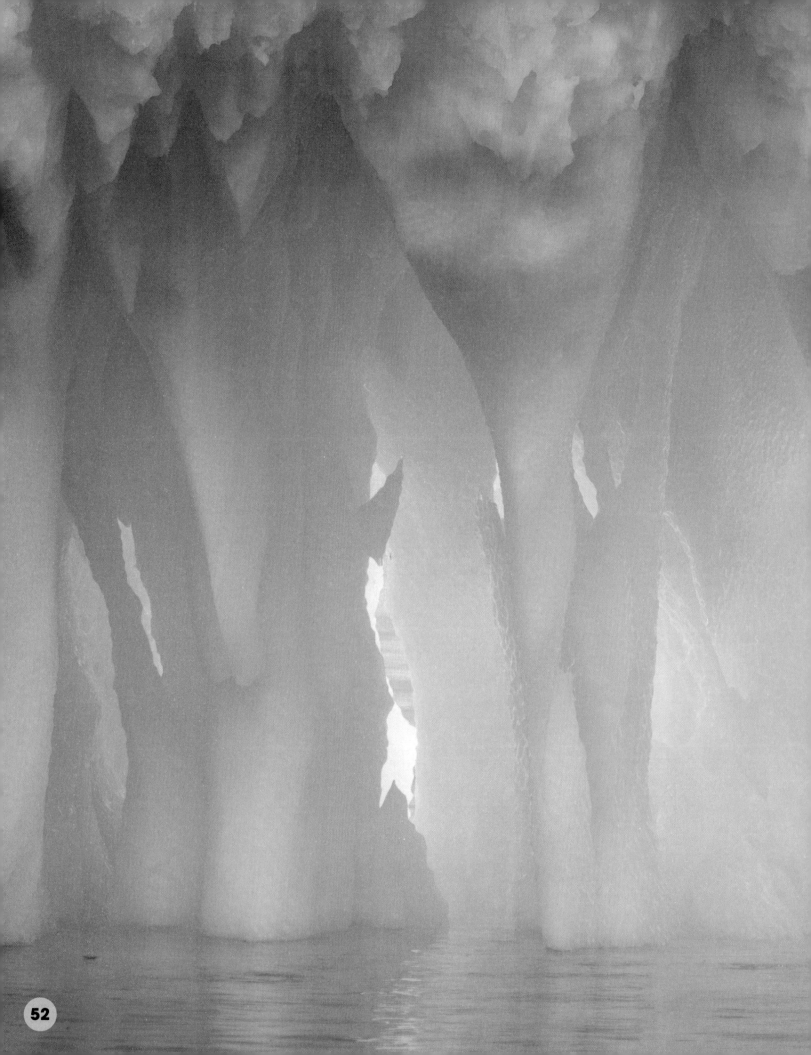

OCEAN HABITATS

The ocean contains lots of different habitats. It is home to thousands of plant and animal species.

The ocean is one huge body of water, but it is not the same all over. Some parts of the ocean are shallow, light, and warm. Other areas are deep, dark, and extremely cold. Deep-sea animals need very special adaptations to survive the harsh conditions found in the deepest, darkest parts of the ocean.

Read on to discover the incredible variety of ocean habitats.

Coral reefs

Coral reefs are brightly coloured and bursting with life. They can be found in warm, tropical waters with lots of light. Reefs are very delicate environments, which are easily damaged by changing temperatures and pollution.

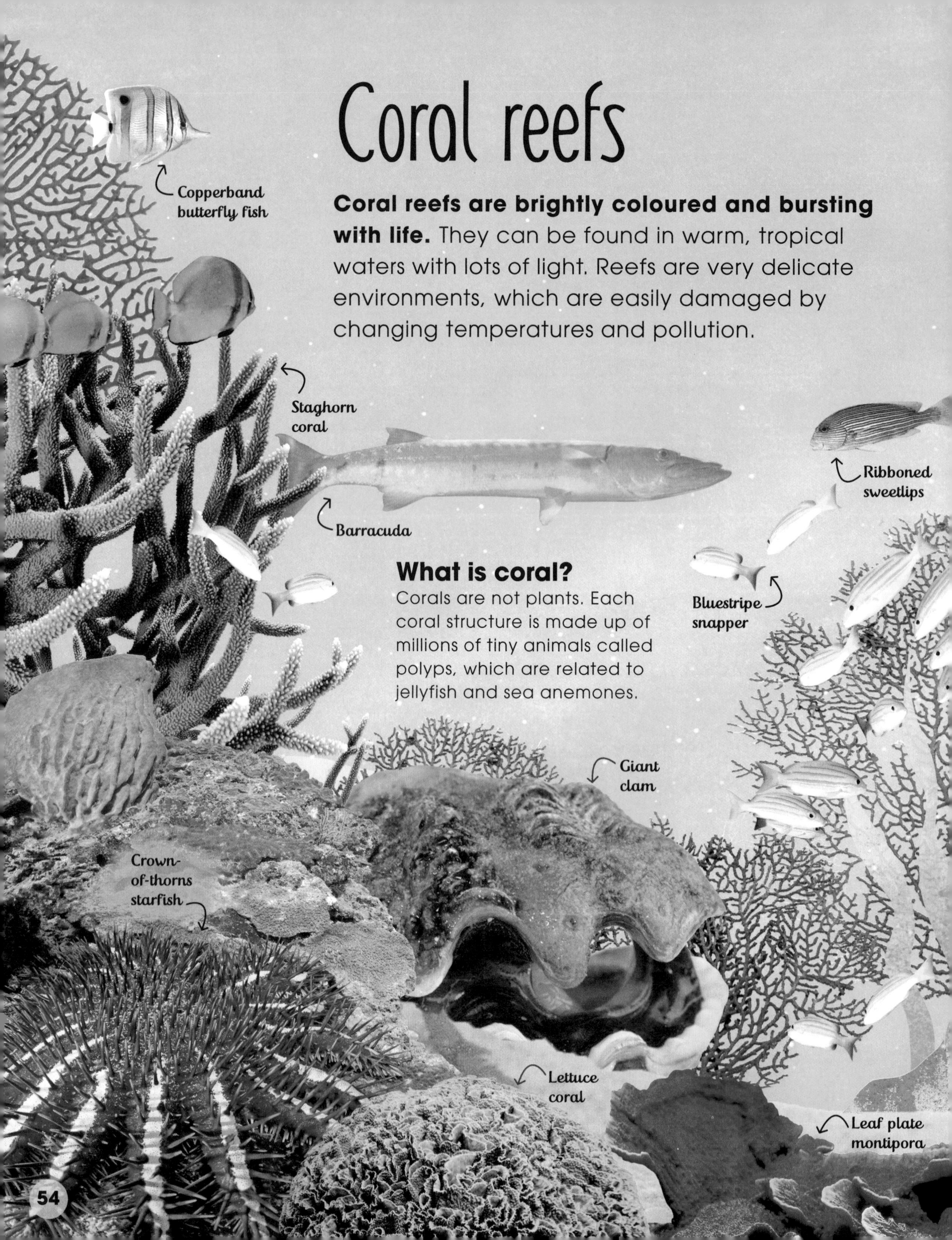

Copperband butterfly fish

Staghorn coral

Barracuda

Ribboned sweetlips

What is coral?

Corals are not plants. Each coral structure is made up of millions of tiny animals called polyps, which are related to jellyfish and sea anemones.

Bluestripe snapper

Giant clam

Crown-of-thorns starfish

Lettuce coral

Leaf plate montipora

Coral reefs are found in less than 1 per cent of the ocean, but are home to 25 per cent of ocean plant and animal species.

Hawksbill turtle

Bumphead parrotfish

Blotcheye soldierfish

Hard or soft?

Hard corals have rock-like skeletons and form the structure of the reef. Soft corals do not have these skeletons and look more like plants.

Walking shark

Eating coral
Adult corals are eaten by fish, worms, and sea stars.

Christmas tree worm

Red whip coral

Sea fan

Pygmy seahorse

Bicolour angelfish

Garden eel

Peacock mantis shrimp

Dragon in disguise
The leafy sea dragon is hard to spot as it swims through the seagrass – its leafy body is the perfect disguise.

Seagrass

Leafy sea dragon

Seagrass meadows

These underwater meadows are found in shallow seas. They are home to hundreds of different types of animals. Big animals visit to eat the seagrass, while smaller ones make hidden homes between the plants.

Longhorn cowfish

Green turtle

Seagrass meadows are found all over the world. Some are so big they can be seen from space!

Dugong

Underwater eating

Dugongs eat seagrass all day and night. They need to breathe air, but can stay underwater for six minutes at a time.

Spotted eagle ray

Blackspot snapper

Fish nursery

Seagrass meadows provide safe shelter for thousands of tiny baby fish. When the fish get bigger, they leave their meadow and head out into the ocean.

Sea cucumber

Sea urchin

Blue sea star

57

Kelp forests

Not far from the coast, thick forests of seaweed provide food and shelter for hundreds of ocean animals, including many that are rare or endangered. These kelp forests also help prevent climate change by taking in carbon dioxide from the air.

Sea otters

Otters protect kelp by eating sea urchins. Sea otters wrap themselves in kelp so they don't float away when they sleep.

Stipe (stalk)

Giant kelp

Anchored to the bed of the ocean, giant kelp reaches towards the surface like a towering underwater tree.

Air bladder
These air-filled pockets help the stalks to stay upright.

Giant kelp is one of the fastest growing things on the planet. It can grow up to 0.6 m (2 ft) a day.

Blacksmith fish

Sea lions

Playful sea lions glide easily through the kelp. They push and chase their way between the seaweed, hunting for fish and other animals to eat.

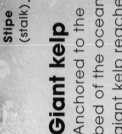

Blade
(leaf)

Garibaldi fish

Colourful crabs

Northern kelp crabs change colour depending on which kelp they eat.

Northern kelp crab

Bat star

Holdfast
This attaches kelp to rocks on the seafloor.

Strawberry anemone

Giant spined sea star

Sea urchins

Sea urchins

Sea urchins love kelp. Normally they eat pieces that fall to the ocean floor. But if they are not kept under control, they can destroy entire kelp forests by chomping through growing plants.

Schools of sardines group into ball shapes when large predators come close.

Gannets

Diving down
Gannets can dive after fish at speeds of up to 97 kph (60 mph).

Sardines

Swordfish

The open ocean

In the pelagic zone there is nothing but water and sea creatures. No seafloor or seashore, just wide open ocean. The Earth's surface is around **70 per cent water,** and the oceans are very deep, which makes the pelagic zone the world's **largest habitat.** Many of the animals that live here are giants.

Krill ball
Swarms of krill form huge,
swirling balls in the ocean.
They are an important source
of food for hundreds of animals,
including squid, fish, seabirds,
and some whales.

Krill

Ocean
sunfish

Barrel
jellyfish

Whale sharks are the biggest fish
in the oceans. They can weigh as
much as nine elephants.

Whale
shark

Along for the ride
Remoras are known as
the hitchhikers of the sea.
Special suction pads on
their heads allow them to
anchor onto the skin of
large ocean animals.

Remora

MARINE SNOW

Flakes of dead material fall downwards from the water's surface. This **"marine snow"** is food for creatures in the deeper waters.

Dumbo octopus

UNUSUAL CREATURES

The unique conditions of the deep ocean mean that the animals here look **totally different** to those found anywhere else.

Deep-sea dwellers

Down in the mysterious depths of the ocean, it is pitch black and freezing cold. It is hard to believe that anything can survive. But lurking in the darkness, **bizarre creatures** thrive.

DEEP-SEA SMOKERS

Many deep-sea creatures rely on the heat and chemicals produced by **underwater hot springs.** The water here can reach 400°C (750°F).

Giant tube worms

Yeti crab

Deep-sea mussels

Spiny crab

Cookiecutter shark

The cookiecutter shark rips **chunks of flesh** from larger animals without killing them.

Whalefish

Gossamer worms

BIOLUMINESCENCE

No sunlight reaches these waters. Instead, the ocean glows with **bioluminescence** – light that is made by the animals that live here.

Anglerfish

The anglerfish uses its light to **attract prey.**

The transparent cockatoo squid uses its lights for **camouflage.**

Cockatoo squid

WHALE FALL

When a whale dies, its body sinks to the bottom of the ocean. Here, it provides **food for hundreds of other animals.** Every part of the carcass is eaten, from the blubber (fat) to the bones.

Bristle worms

Hooded shrimp

Zombie worms

Life under the ice

Huge sheets of frozen fresh water float over the Antarctic coastline. The dark waters beneath them are ghostly quiet and freezing cold. Not many animals can survive in these conditions, but those that do are found in their thousands.

Underwater icicle

When cold brine (salty water) trickles out of the ice sheet and down to the seafloor, it freezes all the water in its path. An **icy column** called a "brinicle" is formed.

Volcano sponge

Icy danger
Any animals that are touched by the brinicle will be frozen to death.

Mega sponge
Volcano sponges can be over 1 m (3 ft) wide and live for thousands of years.

Soft coral

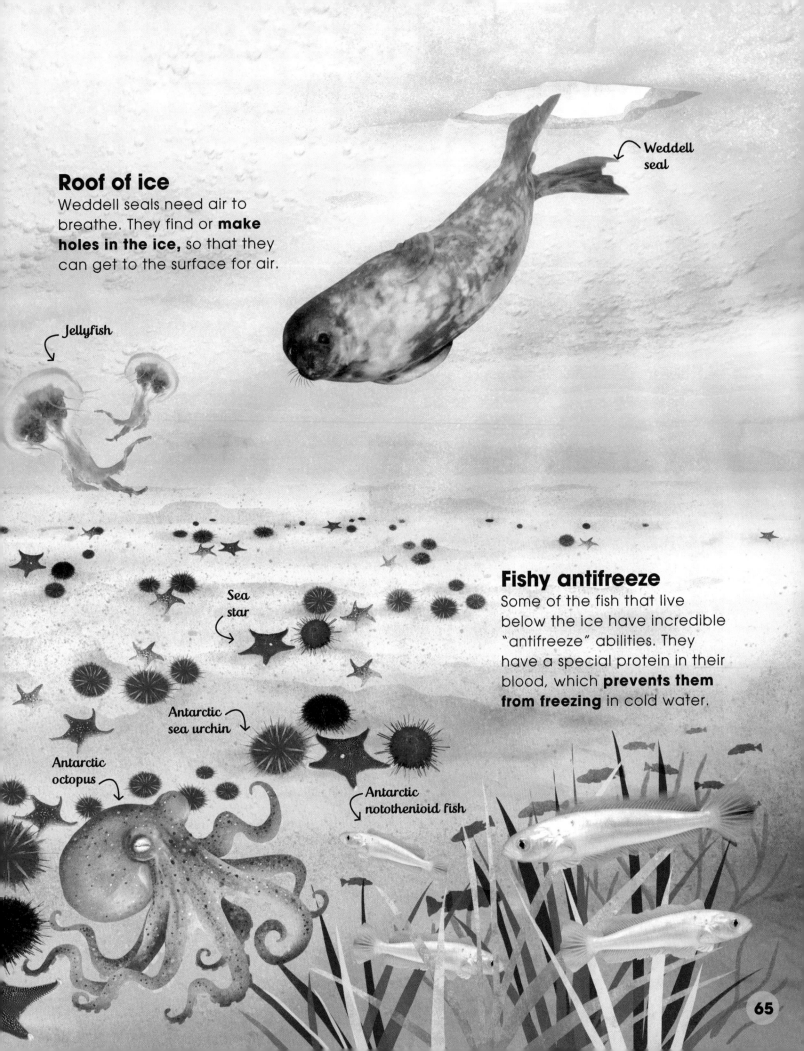

Roof of ice

Weddell seals need air to breathe. They find or **make holes in the ice,** so that they can get to the surface for air.

Weddell seal

Jellyfish

Sea star

Fishy antifreeze

Some of the fish that live below the ice have incredible "antifreeze" abilities. They have a special protein in their blood, which **prevents them from freezing** in cold water.

Antarctic sea urchin

Antarctic octopus

Antarctic notothenioid fish

Seashore

High winds and rough seas make life at the coast difficult. Throughout the day, the water moves up and down the land, its temperature and depth constantly changing. Animals and plants must be able to survive in these shifting conditions.

Breadcrumb sponges

Purple laver

Oystercatcher chick

Oystercatcher

Sea urchin

Danger spikes
Some sea urchins have venom in their spikes.

Cape rock crab

Barehead goby

African black oystercatcher

Rock pools

At low tide on rocky shores, the sea retreats. It leaves behind pools of salty water. The creatures that live in these rock pools are specially adapted to life in and out of the water.

Floating plastic waste often washes up at the coast.

Mussels

Limpets

Hiding away
Small animals shelter in the seaweed.

Klipfish

Sea lettuce

Dwarf cushion-star

Hold tight
Clamping down on rocks stops limpets from being washed away or eaten by birds.

Cream tube sponge

Redbait

OCEANS & ME

Without oceans, our planet would be a completely different place – dry, hot, and lifeless.

Oceans are critical to life on Earth. They provide us with food to eat and oxygen to breathe. They also stop our planet from getting too hot.

Although we rely on the oceans for the normal running of our lives, we don't always treat them very kindly. People are taking trillions of fish from the sea, and a lot of our rubbish ends up in the oceans, killing wildlife.

We must take care of our oceans because we rely on them for the future of people, animals, and our planet.

How oceans help us

The oceans allow us to transport things around the world. They give us food to eat and provide opportunities for adventure. But the true value of oceans is more important than any of these things. Without oceans, life on Earth would simply not exist.

O_2

REMOVE CARBON DIOXIDE

A lot of the carbon dioxide in the air dissolves into the oceans, which helps to stop the planet from warming too quickly.

PRODUCE OXYGEN

Tiny plants called phytoplankton release oxygen into the air. This is great for humans as we need oxygen to breathe!

CO_2

Fun and games

Lots of fun activities involve the oceans, including surfing, snorkelling, sailing, swimming, kayaking, scuba diving, and fishing. It is important to use the oceans safely and respectfully.

Wave and wind power

The power of waves and tides can be used to create energy. Wind turbines are also built at sea. They make energy from the wind.

Weather

The oceans play an important role in keeping our planet warm. They absorb heat from the Sun and spread it around the world.

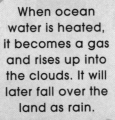

When ocean water is heated, it becomes a gas and rises up into the clouds. It will later fall over the land as rain.

EVAPORATION

Transport

Ocean-going ships transport all sorts of things around the world, from food and building materials to people and animals.

Ocean life

Earth's oceans are home to millions of creatures, which come in many different shapes and sizes.

Taking from the oceans

Many of the things we rely on in our day-to-day lives have been taken from the oceans.

Fish, shellfish, and crustaceans
These animals are all collected for us to eat.

Salt
Seawater is evaporated to harvest its salt.

Seaweed
This ocean plant is used in food and as fuel. It is also used as a fertiliser, to help plants grow.

Medicine
Chemicals extracted from sea creatures, such as cone snails, are used to treat certain diseases.

Fossil fuels
Oil and natural gas are extracted from the seabed.

Changing oceans

We do not always treat our precious oceans with the respect they deserve. Humans have thrown plastic into the oceans, fished without thinking, and caused the whole planet to overheat. We must act soon to protect our oceans, before it is too late.

Plastic at sea
Plastic is found in all layers of the ocean, but it often floats on the surface.

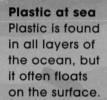

Turtles often eat plastic by mistake.

At least **8 million tons (7.3 million metric tons)** of plastic end up in our oceans every year. Ocean animals are often killed when they get tangled in plastic or eat it.

Hermit crab with plastic "shell".

PLASTIC

Our oceans are full of plastic waste, but we are still dumping tons of it every single day. It takes hundreds of years for plastic to break down into small pieces – and those small pieces are harmful, too.

Hunting and fishing

Humans once believed that there were so many animals in the sea, they would never run out. They were wrong. Many ocean species are now in danger of dying out because they have been hunted and fished for so long.

Marine protected areas

Marine reserves are areas where hunting and collection is not allowed. Creating reserves gives animals and plants in badly affected areas a chance to recover.

Hawaiian monk seal

Tens of millions of sharks are killed for their fins every year. The rest of their bodies are often tossed back into the ocean.

Global warming

As the world heats up, glaciers and ice sheets are melting. This extra liquid has caused sea levels around the world to rise. Rising sea temperatures are also bad for ocean wildlife, such as corals.

Healthy coral
Brightly coloured algae live inside coral and provide it with food.

Stressed coral
Warmer ocean temperatures and pollution make the coral stressed, and it forces the algae out.

Bleached coral
The coral that is left cannot feed itself, so it starves.

Helping oceans

The best way to keep plastic out of the oceans is not to buy it. But if you can't avoid plastic, why not reuse it? You can use unwanted plastic to make something new.

SELF-WATERING PLANT POT

Plastic is a good material to use for planters because it is waterproof and strong. This plant pot waters itself at the same time as saving the ocean.

This is where you will attach the string.

Getting ready

To create this planter, you will need a plastic bottle, scissors, ruler, string, water, soil, and a plant. You will also need an adult to help with the cutting.

First steps

Ask an adult to help you make a small hole in the bottle's lid using scissors. Then carefully cut around your bottle just over halfway up.

Less than 20 per cent of the plastic bottles we use are recycled afterwards.

Be careful of the sharp edges.

The string will allow your plant to drink.

Your plant will grow best in a sunny spot.

Soil

Water

Put it together
Thread the string through the hole in the bottle lid, then replace the lid on top of the bottle. Turn the top part of your bottle over and place it inside the bottom part.

Add your plant
Fill the bottom of your bottle with water and the top of the bottle with soil. Make a hole in the soil to put your plant into, then press the soil firmly down around it.

Glossary

ADAPT
When an animal or plant changes to become better-suited to its habitat.

ALGAE
Simple, plant-like organism that is found in or near water. Seaweed is a type of algae.

ANTARCTIC
Region around the South Pole.

ARCTIC
Region around the North Pole.

BIOLUMINESCENCE
Production of light by a living organism.

BLOWHOLE
Nostril on top of a cetacean's head that allows it to breathe.

BLUBBER
Thick layer of fat that protects a marine mammal from the cold.

CAMOUFLAGE
Usually skin colouring or pattern that helps an animal blend into its surroundings.

CARBON DIOXIDE
Gas in the air that plants absorb and use to make food.

CEPHALOPOD
Type of mollusc, such as an octopus, with jaws and sucker-covered limbs.

CETACEAN
A marine mammal. Group includes whales, dolphins, and porpoises.

CLIMATE
Weather that is typical for a specific area over a long period of time.

CLIMATE CHANGE
Change in temperature and weather across the Earth.

COAST
Where land meets the sea.

COLONY
Large group of one species of animal or plant that lives close together.

CONTINENTAL CRUST
Thick slab of light rock which forms part of the Earth's crust. It moves over the heavier rock of the Earth's mantle and forms the continents.

CONTINENTAL DRIFT
Process by which continents are moved around the Earth.

CONTINENTAL SHELF
Part of a continent that is submerged under a relatively shallow area of water.

CONTINENTAL SLOPE
Edge of the continental shelf which tilts down to the ocean floor.

CORAL
Tiny sea animal with a hard outer skeleton that lives in a colony.

CORAL REEF
Rock-like structure formed by groups of corals in the warm waters along tropical coasts. Many fish and other sea creatures live around coral reefs.

CRUSTACEAN
Animal with jointed legs and a tough, jointed outer skeleton which covers its body, such as a crab, shrimp, or lobster.

CURRENT
The movement of water in a specific direction.

DORSAL FIN
The single fin on the back of an animal, such as a dolphin.

EARTHQUAKE
Sudden movements in the Earth's crust causing the ground to shake violently.

ECHINODERM
Invertebrate with spines in the skin, such as a sea star (starfish).

ENVIRONMENT
The surroundings in which an animal lives, including land and weather.

FOOD WEB
Connections between living organisms that eat each other.

FOSSIL
Remains of an ancient living thing that has been turned into rock.

FOSSIL FUEL
Type of natural energy formed from the remains of ancient animals. They are not renewable (cannot be used again). Oil is a fossil fuel.

FRESH WATER
Water that is not salty, such as rivers and ponds.

GASTROPOD
Soft-bodied creature with tentacles and hundreds of teeth, such as a snail or slug. Part of the biggest group of molluscs.

GILL
Part of a fish or crustacean that allows it to absorb oxygen and breathe underwater.

GLOBAL WARMING
The rise in the Earth's average temperature.

HABITAT
Natural home of plants or animals, such as a seagrass meadow or coral reef.

INVERTEBRATE
Animal without a backbone.

KRILL
Small crustacean of the open seas that is eaten by many larger ocean animals.

LIFE CYCLE
The changes a living thing goes through over its lifetime.

MAMMAL
Warm-blooded vertebrate animal that feeds milk to its young.

MARINE
Describes animals that live in the sea and the underwater environment.

MIGRATION
Movement of animals or people from one place to another. Usually seasonal and for protection, feeding, or breeding.

MOLLUSC
Soft-bodied animal that may have a shell, such as a snail or oyster.

OXYGEN
Gas in the air, which all living things need to live.

PINNIPED
Marine animal with fins for feet. Includes sea lions, seals, and walruses.

PLANKTON
Tiny living things that drift in oceans and lakes, often near the surface.

POISON
Harmful substance that may be deadly if eaten or touched.

POLAR REGION
Area around either the North Pole or the South Pole.

PREDATOR
Animal that hunts and eats other animals.

PREY
Animal that is hunted for food by other animals.

REPTILE
Cold-blooded animal with dry, scaly skin that typically lays eggs on land. Includes sea turtles, crocodiles, and sea snakes.

SCALES
Hard pieces in the skin that cover a reptile's body.

SCHOOL (OF FISH)
Large group of fish that swim together in the same direction and in a coordinated style.

SEA
Part of the ocean that is partly surrounded by land. Seas are smaller areas of salt water than oceans.

SEAMOUNT
Underwater volcano on the ocean floor that is fully submerged.

SEASHORE
Area of sandy, stony, or rocky land that borders the sea.

SEA LEVEL
Average height of the sea when it meets the land.

SHOAL
Large group of fish that swim together.

SONAR
Device that uses sound waves to work out how deep or far away animals or objects are in the water. Sonar is used for navigation and communication.

SPECIES
Groups of animals or plants that breed together and have shared characteristics.

STREAMLINED
Smoothly shaped to move easily through water.

SYMBIOSIS
Relationship between two different species that benefits one or both animals.

TENTACLE
Long, boneless, flexible body part that underwater animals, such as squids and octopuses, use for touching or holding.

TIDE
The regular rise and fall of the ocean, caused by the pull of the Moon.

TRENCH
Steep-sided trough or valley in the ocean floor.

TSUNAMI
Huge wave created by an earthquake or volcanic eruption in the ocean.

VERTEBRATE
Animal with a backbone.

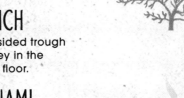

Index

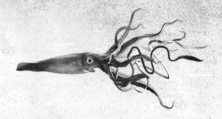

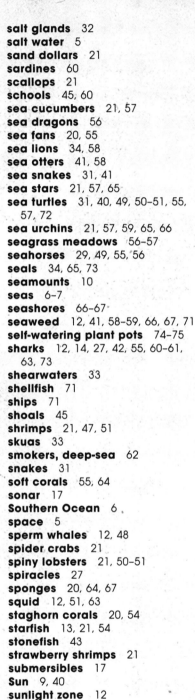

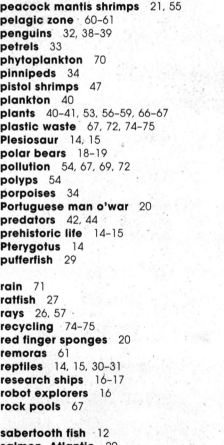

Acknowledgements

The publisher would like to thank the following people for their assistance:
Polly Goodman for proofreading; Helen Peters for the index; Dheeraj Arora for help with jacket finishes; and Tom Morse for CTS help.

PICTURE CREDITS

The publisher would like to thank the following for their kind permission to reproduce their photographs: (Key: a-above; b-below/bottom; c-centre; f-far; l-left; r-right; t-top)

1 123RF.com: Ten Theeralerttham / rawangtak (cb). **Alamy Stock Photo:** WaterFrame (clb/Garibaldi Fish and Giant Spined Starfish). **Dreamstime.com:** Pablo Caridad / Elnavegante (cra); Underwatermau (tc); Dongfan Wang / Tabgac (cla); Cherdchay Toyhem (clb, crb/seaweed); Lgor Dolgov / Id1974 (cb/Echinus esculentus, crb/Echinus esculentus); Fenkie Sumolang / Fenkieandreas (crb); Sombra12 (cr). **Fotolia:** uwimages (cla/anemonefish). **2 Dorling Kindersley:** Linda Pitkin (clb, bl). **2-3 Dreamstime.com:** Designprintck (Background). **4-5 Alamy Stock Photo:** eye35 stock. **5 Dreamstime.com:** Designprintck (t). **6 Dreamstime.com:** Pablo Caridad / Elnavegante (cra). **6-7 Dreamstime.com:** Ruslan Nassyrov / Ruslanchik (t). **8 123RF.com:** Sirapob Konjay (t). **Dreamstime.com:** Pablo Caridad / Elnavegante (ca); Vladvitek (cra). **9 123RF.com:** Steve Collender (tc); Marigranula (crb/palm). **Dreamstime.com:** Mexrix (cb/Sea); Ruslan Nassyrov / Ruslanchik (cb). **Fotolia:** Yong Hian Lim (cr, crb). **10 Dreamstime.com:** Vladvitek (tc). **10-11 Dreamstime.com:** Ruslan Nassyrov / Ruslanchik (c). **12 123RF.com:** Pavlo Vakhrushev / vapi (ftl). **Alamy Stock Photo:** National Geographic Image Collection (cb); Paulo Oliveira (tl, tl/hatchet fish); Nature Picture Library (cla); World History Archive (bc); Kelvin Aitken / VWPics (bc/anglerfish). **Dorling Kindersley:** Holts Gems (fbl); Natural History Museum, London (fclb). **Dreamstime.com:** Caan2gobelow (tl/dolphins); Tazdevilgreg (ftl/coral Trout); Jamesteohart (cla/Whale); Tatus (cla/Chelmon); Carol Buchanan / Cbpix (fcla). **14-15 Dreamstime.com:** Designprintck (Background). **14 Dreamstime.com:** Cornelius20. **15 Dorling Kindersley:** Hunterian Museum University of Glasgow (cl). **Dreamstime.com:** Eugene Sim Junying (crb); Tententenn (cra). **Science Photo Library:** Millard H. Sharp (cr). **16 Alamy Stock Photo:** Agencja Fotograficzna Caro (c); PJF Military Collection (tr). **17 Dreamstime.com:** Kateryna Levchenko (cl); Willyambradberry (cra). **18-19 Alamy Stock Photo:** National Geographic Image Collection. **20 Alamy Stock Photo:** David Fleetham (c); National Geographic Image Collection (cla); Luiz Puntel (cb); imageBROKER (crb). **Dorling Kindersley:** Linda Pitkin (cra, cr/Giant feather duster worm). **Dreamstime.com:** Seadam (clb); WetLizardPhotography (fcla). **Getty Images / iStock:** atese (cr). **naturepl.com:** David Shale (fclb, fclb/Glass sponge). **21 Alamy Stock Photo:** Brandon Cole Marine Photography (cla/Snail). **Dorling Kindersley:** Natural History Museum, London (cl, crb); Linda Pitkin (cla, fclb, tr). **Dreamstime.com:** Aleksey Solodov (cra). **22 Dreamstime.com:** Blufishdesign (cl); Iulianna Est (crb). **23 Dreamstime.com:** Seadam (c). **24-25 Dreamstime.com:** Ihor Smishko (b). **25 Alamy Stock Photo:** imageBROKER (c); WaterFrame (tc); Mike Veitch (bc/octopus, br). **Dreamstime.com:** Izanbar (bc); Jolanta Wojcicka (clb). **26 Dorling Kindersley:** Dr. Peter M Forster (cl). **Getty Images:** Westend61 (cb). **26-27 Dreamstime.com:** Andreykuzmin. **27 Alamy Stock Photo:** Kelvin Aitken / VWPics (bl). **Getty Images:** Dmitry Miroshnikov (t). **28 123RF.com:** Witold Kaszkin (b/winter landscape); Eugene Sergeev (c/ice). **Dreamstime.com:** Christopher Wood / Chriswood44 (b/ice); Olga Khoroshunova (Background). **30-31 Dreamstime.com:** Olga Khoroshunova (Water); Ihor Smishko (ca/Sand). **31 Alamy Stock Photo:** Mauritius Images GmbH (c); Ellen McKnight (b/Leatherback turtle). **Dreamstime.com:** Fenkie Sumolang / Fenkieandreas (t); Mexrix (b). **32 Alamy**

Stock Photo: Accent Alaska.com (bl). **Dreamstime.com:** Willtu (tr). **FLPA:** Terry Whittaker (c). **33 Alamy Stock Photo:** AGAMI Photo Agency (c). **Dorling Kindersley:** George Lin (ca). **Dreamstime.com:** Donyanedomam (br). **Getty Images / iStock:** Henk Bogaard (cra). **34-35 Dreamstime.com:** Designprintck (Background); Mexrix (b). **34 Dreamstime.com:** Willyambradberry (br); Vladimir Melnik / Zanskar (cla). **35 123RF.com:** Christopher Meder / ozbandit (br). **Dreamstime.com:** Jamesteohart (bl); Mexrix (ca). **36-37 123RF.com:** Witold Kaszkin (ca/ice); Eugene Sergeev (ca). **Dreamstime.com:** Christopher Wood / Chriswood44 (ca/Frozen). **36 Alamy Stock Photo:** Corey Ford (c). **naturepl.com:** Doug Allan (bl). **37 Dreamstime.com:** Luis Leamus (tl); Zhykharievavlada (clb). **naturepl.com:** Eric Baccega (ca). **38-39 Alamy Stock Photo:** Giedrius Stakauskas. **39 Dreamstime.com:** Designprintck. **40 123RF.com:** Dmytro Pylypenko (cla). **Dorling Kindersley:** Natural History Museum, London (cr). **Dreamstime.com:** Duncan Noakes (c). **Photolibrary:** Photodisc / White (cb). **40-41 Dreamstime.com:** Designprintck (Background); Mexrix (Sea). **41 Dreamstime.com:** Lukas Blazek (cb); Mauro Rodrigues (tl); Donyanedomam (tr); Cherdchay Toyhem (cla); Jamesteohart (c); Simone Gatterwe / Smgirly (br). **42 123RF.com:** Tudor Antonel Adrian / Tony4urban (cl). **Dreamstime.com:** Orlandin (bl). **43 Dorling Kindersley:** Linda Pitkin (cl). **Dreamstime.com:** Andamanse (cla); Seatraveler (c). **44-45 123RF.com:** inkdrop. **45 Alamy Stock Photo:** Frank Hecker (crb). **Dreamstime.com:** Aquanaut4 (cra); Designprintck (r); Seaphotoart (ca). **46-47 Dreamstime.com:** Designprintck (Background). **47 Alamy Stock Photo:** cbimages (cl); Howard Chew (c); imageBROKER (cb); Pete Niesen (br). **Dreamstime.com:** Jolanta Wojcicka (br/Coral reef). **48 Alamy Stock Photo:** Arco Images GmbH (tl); Blue Planet Archive (bl). **48-49 123RF.com:** Ten Theeralerttham / rawangtak (bc). **Dreamstime.com:** Olga Khoroshunova (water); Mexrix (br). **49 Alamy Stock Photo:** Arco Images GmbH (tc); Nature Picture Library (bl). **Dreamstime.com:** Designprintck (t/texture); Ihor Smishko (t/Sand); Foryouinf (cra). **FLPA:** Photo Researchers (cr). **50 Alamy Stock Photo:** Nature Picture Library (cla). **Dreamstime.com:** Wrangel (crb). **51 Dreamstime.com:** Corey A. Ford / Coreyford (ca); Hadot (clb); Schnapps2012 (cb). **52-53 Alamy Stock Photo:** Ron Niebrugge. **53 Dreamstime.com:** Designprintck. **54-55 123RF.com:** Ten Theeralerttham / rawangtak (cb). **Dreamstime.com:** Daisuke Kurashima. **54 Alamy Stock Photo:** F1online digitale Bildagentur GmbH (bc); Luiz Puntel (cl); WaterFrame (bl); Reef and Aquarium Photography (br). **Dorling Kindersley:** Linda Pitkin (ca). **Dreamstime.com:** Orlandin (cra); Secondshot (ca). **Getty Images / iStock:** marrio31 (tl). **55 123RF.com:** sergemi (cb). **Alamy Stock Photo:** imageBROKER (br); Oksana Maksymova (cra). **Dorling Kindersley:** Holts Gems (cra/Gem Coral); Linda Pitkin (ca, fcrb); Natural History Museum, London (clb/coral branch). **Dreamstime.com:** Alexander Shalamov / Alexshalamov (tr); Whitcomberd (bc); Orlandin (cr, bl); Ethan Daniels (cr/Shdrk); Dream69 (fcra, bc/Gorgon whip). **Getty Images / iStock:** Peter_Horvath (clb). **56 Dreamstime.com:** Juliana Scoggins / Bellavista233 (ca); Serg_dibrova (bl); Seadam (bc); Isselee (crb). **57 123RF.com:** Richard Carey (tl); Keith Levit (c). **Alamy Stock Photo:** Martin Strmiska (ca). **Dreamstime.com:** Serg_dibrova (br); Seadam (bl); Para827 (cb). **60 Alamy Stock Photo:** BIOSPHOTO (ca); Nature Picture Library (tc); Paulo Oliveira (tl). **60-61 Dreamstime.com:** Martin Voeller (b). **61 123RF.com:** aoldman (bc). **Alamy Stock Photo:** Nature Picture Library (tr); David Priddis (cla). **naturepl.com:** Richard Herrmann (tl). **62 123RF.com:** Charles Brutlag (c). **64-65 123RF.com:** Witold Kaszkin (t/ice). **Alamy Stock Photo:** Justin Hofman (c). **Dreamstime.com:** Tarpan. **naturepl.com:** Pascal Kobeh (bc/starfish); Norbert Wu (Sea Urchin). **64 naturepl.com:** Pascal Kobeh (crb); Norbert Wu (b). **65 naturepl.com:** Jordi Chias (cl); Norbert Wu (t, br, crb, cb/Bald Notothen, bc); Pascal Kobeh (cb, cb/starfish). **66 Alamy Stock Photo:** Richard Mittleman / Gon2Foto (c). **Dorling Kindersley:** Cecil Williamson Collection (cb/stone); Stephen Oliver. **Dreamstime.com:** Lgor Dolgov / Id1974 (cb). **naturepl.com:** Richard Du Toit (ca); Ann & Steve Toon (cla). **Shutterstock.com:** Simon_g (b). **66-67 Dreamstime.com:** Steveheap (ca/Water). **naturepl.com:** Hougaard Malan (ca). **67 Alamy Stock Photo:** Martin Lindsay (tc). **Dorling Kindersley:** Cecil Williamson Collection. **Dreamstime.com:** Lgor Dolgov / Id1974 (cb); Alfio Scisetti / Scisettialflo (c); Madelein

Wolfaardt (clb). **naturepl.com:** Nick Upton (bl). **68-69 Getty Images:** Klaus Vedfelt (c). **69 Dreamstime.com:** Designprintck (r). **70-71 Dreamstime.com:** Designprintck (Background); Mexrix (Sea). **71 123RF.com:** petkov (cra). **Dreamstime.com:** Aleksey Bakaleev / Bakalusha (cl); Pablo Caridad / Elnavegante (cb); Winai Tepsuttinun (br). **72 123RF.com:** Aleksey Poprugin (c/plastic bag). **Alamy Stock Photo:** cbimages (bl). **Dorling Kindersley:** Quinn Glass, Britvic, Fentimans (cr). **Dreamstime.com:** BY (c); Alfio Scisetti / Scisettialflo (c/bottles); Indigolotos (c/plastic bottle). **Getty Images / iStock:** Picsfive. **73 Alamy Stock Photo:** ArteSub (cra). **74-75 Dreamstime.com:** Mexrix (c/Sea). **75 Dreamstime.com:** Anton Starikov (b). **76-77 Dreamstime.com:** Designprintck (Background); Olga Khoroshunova (b/water). **78 Dreamstime.com:** Cynoclub (tr). **78-79 Dreamstime.com:** Designprintck (Background). **80 Dreamstime.com:** Designprintck.

Cover images: Front: **Dorling Kindersley:** Stephen Oliver cra, cl; **Dreamstime.com:** Cynoclub bl, Fenkie Sumolang / Fenkieandreas bc, Martinlisner ca/ (Blue Tang), Sombra12 cr, Tazdevilgreg ca; Back: **Dorling Kindersley:** Stephen Oliver clb; **Dreamstime.com:** Cynoclub ca, Dragonimages tl, Sombra12 cl; Spine: **Dreamstime.com:** Sombra12 cb

Endpaper images: Front: **Dreamstime.com:** Cynoclub ; Back: **Dreamstime.com:** Cynoclub

All other images © Dorling Kindersley
For further information see: www.dkimages.com

ABOUT THE ILLUSTRATOR

Claire McElfatrick is a freelance artist. She created illustrated greetings cards before working on children's books. Her beautiful hand-drawn and collaged illustrations are inspired by her home in rural England.

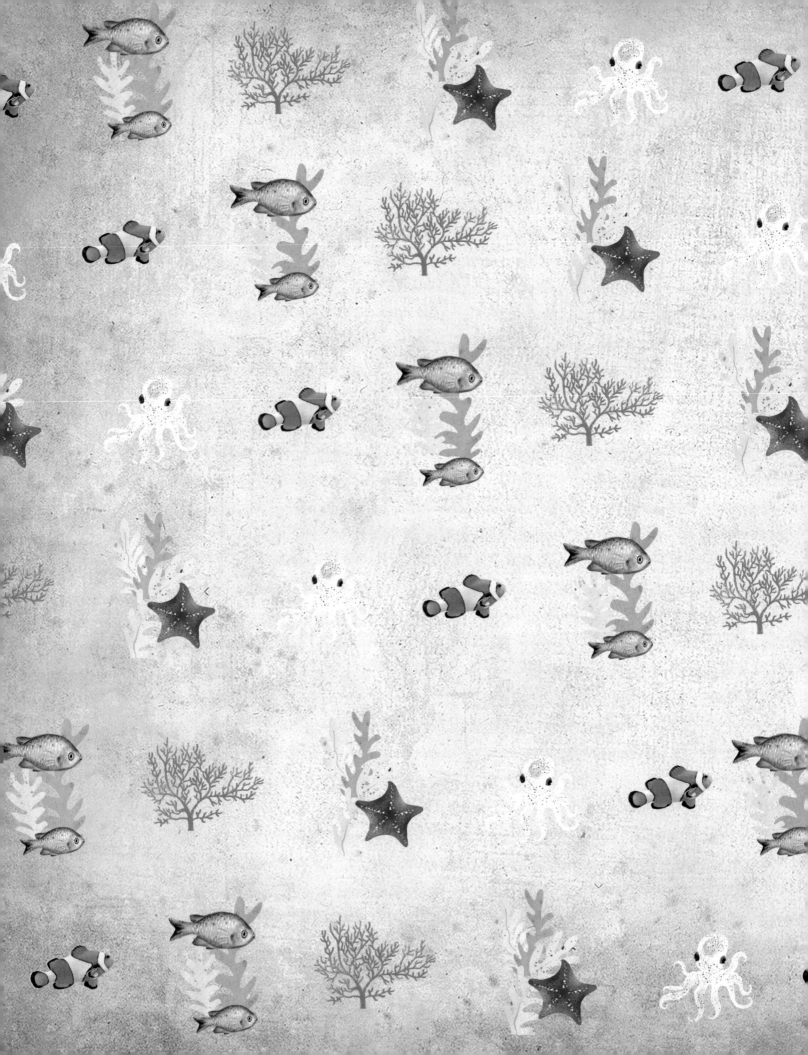